COLLINS

Student Support Materials for

AQA

AS PHYSICS

SPECIFICATION (A)

Module 1: **Particles, Radiation and Quantum Phenomena**

Dave Kelly
Series editor: John Avison

This booklet has been designed to support the AQA (A) Physics AS specification. It contains some material which has been added in order to clarify the specification. The examination will be limited to material set out in the specification document.

Published by HarperCollins*Publishers* Limited
77–85 Fulham Palace Road
Hammersmith
London
W6 8JB

www.**Collins**Education.com
Online support for schools and colleges

© HarperCollins*Publishers* Limited 2000
First published 2000, Reprinted 2001, 2002, 2003

ISBN 0 00 327715 1

Dave Kelly asserts the moral right to be identified as the author of this work

All rights reserved. No part of this publication may be reproduced, stored in a retrieval system, or transmitted in any form or by any means, electronic, mechanical, photocopying, recording or otherwise, without either the prior permission of the Publisher or a licence permitting restricted copying in the United Kingdom issued by the Copyright Licensing Agency Ltd., 90 Tottenham Court Road, London W1P 0LP.

British Library Cataloguing in Publication Data
A catalogue record for this publication is available from the British Library

Cover designed by Chi Leung
Editorial, design and production by Gecko Limited, Cambridge
Printed and bound by Scotprint

The publisher wishes to thank the Assessment and Qualifications Alliance for permission to reproduce the examination questions.

You might also like to visit

www.harpercollins.co.uk
The book lover's website

Other useful texts

Full colour textbooks
Collins Advanced Modular Sciences: Physics AS
Collins Advanced Science: Physics

Student Support Booklets
AQA (A) Physics: 2 Mechanics and Molecular Kinetic Theory
AQA (A) Physics: 3 Current Electricity and Elastic Properties of Solids

To the student

What books do I need to study this course?

You will probably use a range of resources during your course. Some will be produced by the centre where you are studying, some by a commercial publisher and others may be borrowed from libraries or study centres. Different resources have different uses – but remember, owning a book is not enough – it must be *used*.

What does this booklet cover?

This *Student Support Booklet* covers the content you need to know and understand to pass the module test for AQA (A) Physics Module 1: Particles, radiation and quantum phenomena. It is very concise and you will need to study it carefully to make sure you can remember all of the material.

How can I remember all this material?

Reading the booklet is an essential first step – but reading by itself is not a good way to get stuff into your memory. If you have bought the booklet and can write on it, you could try the following techniques to help you to memorise the material:

- underline or highlight the most important words in every paragraph
- underline or highlight scientific jargon – write a note of the meaning in the margin if you are unsure
- remember the number of items in a list – then you can tell if you have forgotten one when you try to remember it later
- tick sections when you are sure you know them – and then concentrate on the sections you do not yet know.

How can I check my progress?

The module test at the end is a useful check on your progress – you may want to wait until you have nearly completed the module and use it as a mock exam or try questions one by one as you progress. The answers show you how much you need to do to get the marks.

What if I get stuck?

A colour textbook such as *Collins Advanced Modular Sciences: Physics AS* provides more explanation than this booklet. It may help you to make progress if you get stuck.

Any other good advice?

- You will not learn well if you are tired or stressed. Set aside time for work (and play!) and try to stick to it.
- Don't leave everything until the last minute – whatever your friends may tell you it doesn't work.
- You are most effective if you work hard for shorter periods of time and then take a (short!) break. 30 minutes of work followed by a five or ten minute break is a useful pattern. Then get back to work.
- Some people work better in the morning, some in the evening. Find out which works better for you and do that whenever possible.
- Do not suffer in silence – ask friends and your teacher for help.
- Stay calm, enjoy it and … good luck!

There are rigorous definitions of the main terms used in your examination – memorise these exactly.

The examiner's notes are always useful – make sure you read them because they will help with your module test.

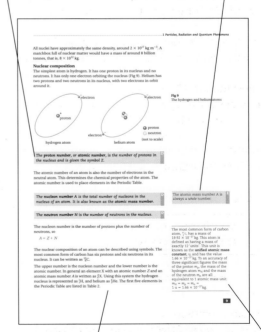

The main text gives a very concise explanation of the ideas in your course. You must study all of it – none is spare or not needed.

Further explanation references give a little extra detail, or direct you to other texts if you need more help or need to read around a topic.

10.1 Particles

10.1.1 Constituents of the atom

The electron

The electron was first identified during experiments using electrical discharge tubes (Fig 1). When the voltage is turned on, the screen at the end of the tube emits a glow. The glow was said to be caused by 'cathode rays'. When the rays hit the screen, their energy is converted into light. This energy conversion, known as **fluorescence**, is aided by coating the inside of the screen with a phosphor, such as zinc sulphide.

Fig 1 An electrical discharge tube

In 1897, J. J. Thomson discovered that he could deflect the rays using electric or magnetic fields (Fig 2). He balanced the two deflections so that the rays moved in a straight line. This allowed him to calculate the charge–mass ratio for the particles making up the rays. He concluded that cathode rays were tiny, negatively charged particles, now called **electrons**.

Electrons are thought to be fundamental particles; there is no evidence that the electron can be broken down into any other particles.

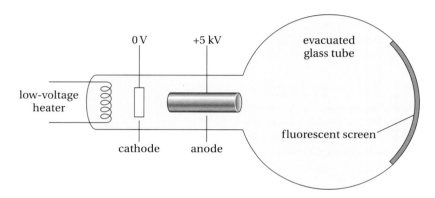

Fig 2 Electrostatic and magnetic deflection: (a) cathode rays curve towards the positive plate, showing they carry a negative charge; (b) a magnetic field tends to make the cathode rays move in a circular path

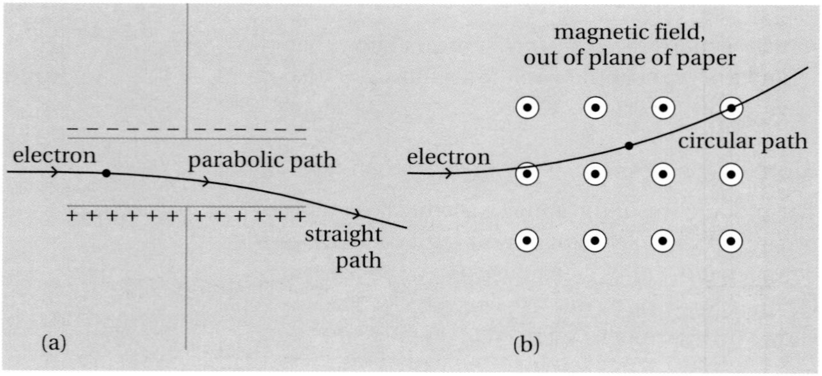

> **E** The force on the electrons acts from the negative cathode to the positive anode: the reverse of the direction of the electric field. The force on the electrons in a magnetic field is at right angles to the field and the motion of the electrons.

Electrons are emitted from the negative electrode, or **cathode**. More electrons are emitted if the cathode is heated. This effect is known as **thermionic emission**. After they are emitted the electrons accelerate towards the anode, finally hitting the screen, where they cause fluorescence.

Thomson realised that the electron was torn away from the rest of the atom by the electrical potential. He suggested that atoms were composed of many electrons, moving in various orbits inside a positively charged cloud. This model of atomic structure is often referred to as the 'plum pudding' model of the atom (Fig 3).

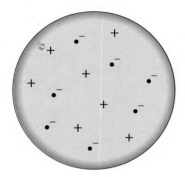

Fig 3
Simple picture of 'plum pudding' model of atom

The charge on an electron is very small. One coulomb is the amount of charge that flows past a point when 1 ampere of current flows for one second. If this was all carried by electrons, there would be 6.25×10^{18} electrons flowing past each second.

Properties of the electron

Mass of the electron $m_e = 9.1 \times 10^{-31}$ kg

Charge of the electron $e = 1.6 \times 10^{-19}$ C

Charge–mass ratio $\dfrac{e}{m_e} = 1.76 \times 10^{11}$ C kg^{-1}

The nuclear atom

In our present model, the atom has a central nucleus where all the positive charge and nearly all the mass are located. Tiny, negatively charged electrons orbit this nucleus, rather like planets orbiting the Sun (Fig 4).

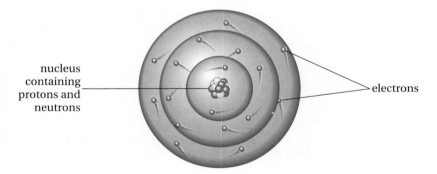

Fig 4
Nuclear atom

nucleus containing protons and neutrons

electrons

Inside the nucleus

Protons *are the particles that carry the positive charge in the nucleus.*

In a neutral atom, that is an atom with no net charge, the number of protons in the nucleus is balanced by the number of electrons orbiting the nucleus. A hydrogen atom has one proton and one electron. Helium has two protons and two electrons, and so on through the Periodic Table, until the heaviest naturally occurring element, uranium, which has 92 protons and 92 electrons.

What holds the nucleus together? Positive charges repel each other and at such short distances the electrostatic forces pushing the nucleus apart are very large. Another force acts inside the nucleus, known as the **strong nuclear force** or **strong interaction**. The strong nuclear force has a very short range; it has no effect at separations greater than about 5 fm (5 × 10^{-15} m) (Fig 5). The strong interaction is an attractive force until the separation is less than 1 fm, when the force becomes strongly repulsive. The overall effect of the strong interaction is to pull the nucleus together, but the repulsive action prevents it from collapsing to a point.

Fig 5
Force–distance graph for strong nuclear force

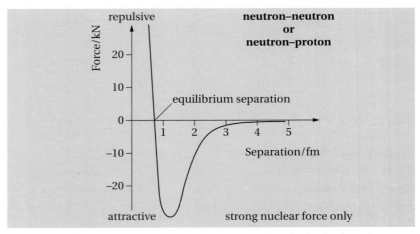

Fig 6
Force–distance graph for proton–proton pair (strong nuclear force plus electrostatic force)

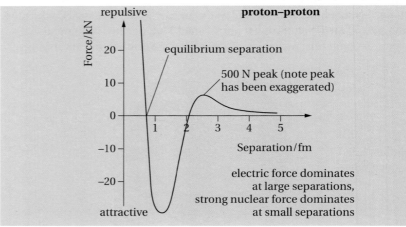

At distances of less than about 2 fm, the strong nuclear attraction between two protons is larger than the electrostatic repulsion (Fig 6) so the nucleus is held together.

For larger nuclei there is a problem. The strong nuclear force acts over a much shorter range than electrostatic repulsion. It isn't possible to get all the protons close enough together for the strong nuclear force to overcome the electrostatic repulsion. There has to be some other particle in the nucleus that helps to glue it all together. This is the **neutron**, discovered by James Chadwick in 1932.

> **D** *The **neutron** is a particle with a mass almost identical to that of the proton, but with no electric charge.*

The neutron exerts a strong nuclear attraction on protons and on other neutrons. Protons and neutrons are the only particles in the nucleus. They are referred to as **nucleons**. The strong nuclear force acts between *any* pair of nucleons, whether that is two protons, two neutrons or a proton and a neutron. Electrostatic repulsion acts only between protons.

The properties of protons, neutrons and electrons are summarised in Table 1.

	Proton	**Neutron**	**Electron**
symbol	$^{1}_{1}p$	$^{1}_{0}n$	$^{0}_{-1}e$
charge (C)	$+1.602 \times 10^{-19}$	0	-1.602×10^{-19}
mass (kg)	1.6726×10^{-27}	1.6749×10^{-27}	9.109×10^{-31}

Table 1
Comparison of protons, neutrons and electrons

10.1.2 Evidence for existence of the nucleus, qualitative study of Rutherford scattering

Geiger and Marsden, working in Ernest Rutherford's laboratories at Manchester University in 1909, studied the scattering of alpha particles as they passed through a thin piece of gold foil (Fig 7). **Alpha particles** are relatively massive, positively charged particles emitted by some radioactive materials. Geiger and Marsden used a scintillator to detect the alpha particles. (A scintillator is a zinc sulphide screen which emits light whenever an alpha particle strikes it.) The scintillator was observed through a small microscope in a darkened room.

Alpha particles are two protons and two neutrons bound tightly together, the same configuration as a helium nucleus. An alpha particle therefore carries a positive charge which is twice the size of the charge on the electron. The mass of an alpha particle is about 8000 times the electron's mass.

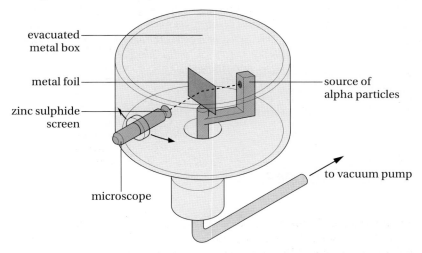

Fig 7
Geiger and Marsden's apparatus

Geiger and Marsden expected to see the alpha particles deflected by small angles. Rutherford suggested that they move the detector in front of the foil, to see if any of the alpha particles were bounced from the surface of the foil. Amazingly, some were; about 1 in every 8000 alphas was 'reflected', or scattered, through an angle of more than 90°.

As an alpha particle travelling at around 10 000 km s^{-1} could not be bounced back by a positively charged mist with tiny electrons embedded in it, Rutherford concluded that almost all the mass of the atom must be gathered together in one small volume, which he called the **nucleus**.

He suggested that the electrons carry all the negative charge and that they orbit the nucleus through empty space a relatively long way from the nucleus. Most of the alpha particles passed through the gold foil with small or zero deflections because they were too far away from the nucleus to be affected by it. Very occasionally an alpha particle would pass so close to the nucleus that it would be repelled by its positive charge and suffer a large deflection (Fig 8).

E — Note that the alpha particles bombard the foil with a lot of energy – turning them round and sending them back requires a very strong force!

Fig 8
Rutherford scattering, showing several paths

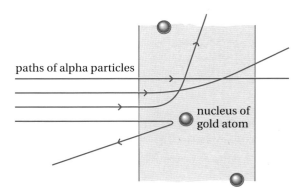

E — When you are sketching the path of a scattered alpha particle it must be nearly in line with a nucleus, and its inbound and rebound paths are symmetrical (a hyperbola).

Analysis of the results allowed Rutherford to work out the scale of the nuclear atom. Whereas the radius of the atom was about 10^{-10} m, the nucleus was only about 10^{-15} m across.

Remember that density, ρ, is defined as the mass in a unit volume:
$$\rho = \frac{m}{v}$$

Example

An atom of gold has 79 electrons orbiting its nucleus. The mass of a gold atom is 3.27×10^{-25} kg. The radius of a gold atom is 144×10^{-12} m. The radius of a gold nucleus is 6.5×10^{-15} m. Find the average density of an atom of gold and the density of a gold nucleus.

Answer

If we assume that the atom is spherical,

Volume = $\frac{4}{3}\pi r^3$

Density = $\frac{mass}{volume}$

So the average atomic density is

$$\frac{3.27 \times 10^{-25} \text{ kg}}{\frac{4}{3}\pi(144 \times 10^{-12} \text{ m})^3} = 26 \times 10^3 \text{ kg m}^{-3}$$

For the nucleus we must subtract the mass of 79 electrons.

Mass of nucleus = 3.27×10^{-25} kg $- 79 \times 9.11 \times 10^{-31}$ kg

= 3.27×10^{-25} kg

Subtracting the mass of the electrons makes very little difference. The mass of the electrons is only about 0.02% of the total mass of the atom.

Density of the nucleus = $\dfrac{3.27 \times 10^{-25}}{\frac{4}{3}\pi(6.5 \times 10^{-15})^3}$

= 2.84×10^{17} kg m^{-3}

For gold, the density of the nucleus is around 10^{13} times higher than the average atomic density.

All nuclei have approximately the same density, around 2×10^{17} kg m^{-3}. A matchbox full of nuclear matter would have a mass of around 8 billion tonnes, 8×10^{12} kg.

Nuclear composition

The simplest atom is hydrogen. It has one proton in its nucleus and no neutrons. It has only one electron orbiting the nucleus (Fig 9). Helium has two protons and two neutrons in its nucleus, with two electrons in orbit around it.

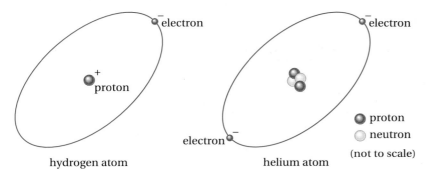

Fig 9
The hydrogen and helium atoms

> The **proton number**, or **atomic number**, is the number of protons in the nucleus and is given the symbol Z.

The atomic number of an atom is also the number of electrons in the neutral atom. This determines the chemical properties of the atom. The atomic number is used to place elements in the Periodic Table.

> The **nucleon number** A is the total number of nucleons in the nucleus of an atom. It is also known as the **atomic mass number**.

The atomic mass number A is always a *whole* number.

> The **neutron number** N is the number of neutrons in the nucleus.

The nucleon number is the number of protons plus the number of neutrons, so

$A = Z + N$

The nuclear composition of an atom can be described using symbols. The most common form of carbon has six protons and six neutrons in its nucleus. It can be written as $^{12}_{6}$C.

The upper number is the nucleon number and the lower number is the atomic number. In general an element X with an atomic number Z and an atomic mass number A is written as $^{A}_{Z}$X. Using this system the hydrogen nucleus is represented as $^{1}_{1}$H, and helium as $^{4}_{2}$He. The first five elements in the Periodic Table are listed in Table 2.

The most common form of carbon atom, $^{12}_{6}$C, has a mass of 19.92×10^{-27} kg. This atom is defined as having a mass of exactly 12 'units'. This unit is known as the **unified atomic mass constant**, u, and has the value 1.66×10^{-27} kg. To an accuracy of three significant figures the mass of the proton m_p, the mass of the hydrogen atom m_H and the mass of the neutron m_n are all equivalent to 1 atomic mass unit:
$m_H = m_p = m_n =$
$1 \text{ u} = 1.66 \times 10^{-27}$ kg.

Table 2
The first five elements in the Periodic Table

Element	Symbol	Atomic number, Z	Neutron number, N	Atomic mass number, A
hydrogen	H	1	0	1
helium	He	2	2	4
lithium	Li	3	4	7
beryllium	Be	4	5	9
boron	B	5	6	11

Isotopes

D *Isotopes are forms of an element with the same proton number but different nucleon numbers.*

Hydrogen usually has one proton, only, in its nucleus, but some hydrogen atoms have one or two neutrons as well (Fig 10).

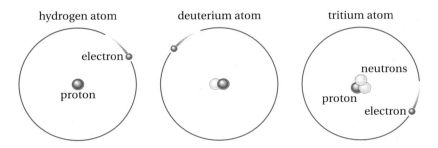

Fig 10 Isotopes of hydrogen. The extra neutrons do not affect hydrogen's chemical behaviour; for example, all three isotopes combine with oxygen to make water

E Isotopes are always the same element. For two isotopes, Z is the same, N and A are different.

The different isotopes of an element have identical chemical behaviour because their atoms have the same number of electrons. Isotopes also have the same number of protons in their nucleus. The difference is simply in the number of neutrons. This makes some isotopes heavier than others. The most common form of carbon has six protons and six neutrons in its nucleus; this isotope is referred to as carbon-12 (Table 3). Carbon-13 has six protons and seven neutrons; carbon-14 has six protons and eight neutrons.

Table 3
Isotopes of carbon

Isotope	Atomic number, Z	Number of electrons	Neutron number, N	Mass number, A	% abundance
carbon-12	6	6	6	12	98.89
carbon-13	6	6	7	13	1.11
carbon-14	6	6	8	14	< 0.001%

10.1.3 Particles, antiparticles and photons

Antimatter

The British physicist Paul Dirac predicted the existence of a particle with exactly the same mass as the electron, but with a positive charge. Indeed he suggested that all particles must have antiparticles.

> An **antiparticle** is a 'mirror image' of a particle, of identical mass but opposite charge.

The first antiparticle was discovered in 1932 by Anderson, who was observing tracks made by cosmic rays, which are high-energy particles from outer space. He was able to see the tracks of high-energy electrons using a cloud chamber. Anderson used a strong magnetic field to curve the path of these electrons. Some tracks seemed identical to the electron tracks but curved in the opposite direction. These tracks were due to an antielectron, now known as a **positron**. A positron is an example of **antimatter**.

> The **positron** is the electron's antiparticle.

When a particle meets its antimatter twin, the particles are drawn together by electrostatic attraction until they annihilate each other (Fig 11).

This is an example of mass–energy equivalence, as predicted by Einstein's Special Theory of Relativity.

> **Annihilation** is the conversion of the mass of a particle and its antiparticle to a pair of photons of electromagnetic radiation.

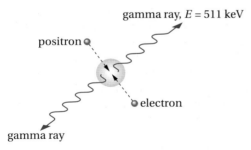

Fig 11 Positron annihilation

When a positron and an electron meet, they annihilate each other. Two identical gamma rays of energy 511 keV are emitted in opposite directions. The electronvolt, eV, is a unit of energy. $1 \text{ eV} = 1.6 \times 10^{-19}$ J, 1 keV = 1000 eV.

The idea that mass is not created or destroyed has to be broadened so that it is part of the *principle of conservation of energy*. The total mass–energy in any system is conserved, but energy and mass may be converted from one to the other. The conversion of mass to energy powers radioactivity and nuclear fission.

Annihilation is the conversion of matter to energy. The opposite process, where matter is 'created' from energy, is called **pair production** (Fig 12).

Fig 12
Pair production: electrons and positrons spiral in opposite directions in the magnetic field

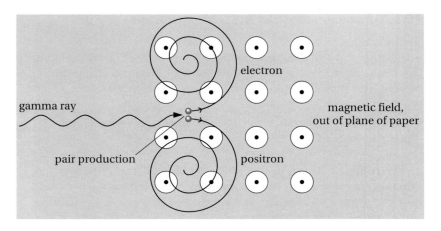

> **D** *Pair production is the process in which a photon of electromagnetic energy is converted to a pair of particles.*

In pair production, there are always two particles created; one is a conventional particle and the other is its antimatter twin. This satisfies the conservation of charge, since before the event there is only a photon of radiation, which carries no charge. After the pair production there are always two particles of opposite charge, making the total charge zero.

A gamma ray has to have a minimum energy of 1.02 MeV before it can create an electron–positron pair. This is because the mass of the pair has an energy equivalent to 1.02 MeV. If the photon has more energy than this, the surplus energy appears as kinetic energy carried by the positron and electron.

In 1955 the first antinucleon was discovered. Protons were accelerated to an energy of up to 6 MeV and collided into other protons in a stationary target. The two protons (p) collided and produced antiprotons ($\bar{p}$) by the reaction:

$$p + p \rightarrow p + p + p + \bar{p}$$

By colliding two protons together we have produced an extra proton and an antiproton. The extra mass needed to create the proton–antiproton pair has come from the kinetic energy of the initial protons.

A year later the antineutron was produced by using antiprotons to collide with protons:

$$\bar{p} + p \rightarrow n + \bar{n}$$

The mass of a particle is always identical to that of its antiparticle, but other properties, such as charge, are opposite.

Neutrinos

Neutrinos are probably the most numerous particles in the universe. They outnumber the protons and neutrons of ordinary matter by a factor of 10^9. Neutrinos created at the time of the Big Bang still permeate the Universe, about 100 or so of them in each cubic centimetre of space. Neutrinos are also emitted by radioactive nuclei and from nuclear reactions. The Earth is bathed in neutrinos from the Sun. Every second about 60 thousand million

> **E** Particles cannot be created unless there is enough energy. The minimum energy needed to create a particle is known as its **rest energy**.

solar neutrinos pass through every square centimetre of the Earth's surface.

Despite this, neutrinos and antineutrinos are extremely difficult to detect. They are not charged. Neutrinos interact with other matter very weakly. Experiments to find neutrinos often use large tanks of water, usually placed well underground, surrounded by sensitive light detectors looking for the occasional flash of light that signifies that a neutrino has interacted with a neutron, or an antineutrino with a proton (see Fig 18).

The neutrino was first predicted by Wolfgang Pauli in 1930. At the time physicists were trying to understand beta radiation (fast-moving electrons emitted by the nuclei of some radioactive atoms). Unlike alpha particles, which are emitted with a well-defined energy, beta particles are emitted with a range of energies (Fig 13). This seemed to contravene the principle of conservation of energy. If a certain amount of energy is transferred by each radioactive decay, why did the emitted beta particle have a range of possible energies? Pauli suggested that another particle, the neutrino, is also emitted in beta decay. The neutrino carries away the balance of the energy, so that the total energy of the decay is always constant.

Neutrinos are so difficult to detect that it wasn't until 1956 that the neutrino was observed experimentally, 26 years after Pauli had predicted its existence.

Beta particles are emitted when a neutron decays into a proton and an electron. The proton stays inside the nucleus but the electron is emitted at high speed. An electron antineutrino is also emitted:
$$n \rightarrow p + e^- + \bar{\nu}_e$$

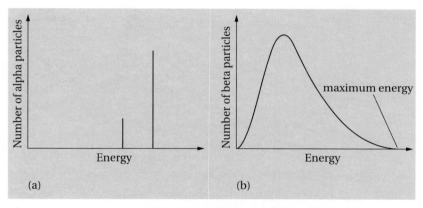

Fig 13
(a) Typical alpha spectrum;
(b) typical beta spectrum

The neutrino is represented by the symbol ν_e and the antineutrino is represented by the symbol $\bar{\nu}_e$. The subscript 'e' stands for 'electron'; these neutrinos are more properly referred to as electron neutrinos, because other types of neutrinos exist.

Some radioactive decays emit a positive beta particle, or positron. This involves the decay of a proton and can be written
$$p \rightarrow n + e^+ + \nu_e$$

The neutrino is believed to be a fundamental particle which carries no charge. For some years the neutrino was believed to have zero mass, but recent experiments (1998) suggest that neutrinos have a small mass, much less than that of an electron.

> The **neutrino** is a fundamental particle with no charge. It has a very small, or zero, mass. It interacts with other matter very weakly.

Photon model of electromagnetic radiation

The wave theory of light is extremely successful in explaining diffraction and refraction. However, there are some phenomena, such as black-body radiation and the photoelectric effect, which cannot be explained by assuming that electromagnetic radiation is composed of waves.

Black-body radiation curves

All objects emit radiation because of their thermal energy. The spectrum of radiation that is emitted depends on the surface temperature of the object and on the type of surface. Objects which are ideal radiators of energy are known as **black-body radiators**.

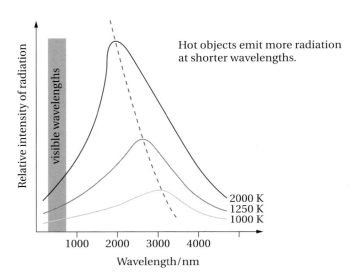

Fig 14
Black-body radiation curves

A perfect black body is an object that absorbs all the radiation that falls on it and reflects none. If this body is in thermal equilibrium with its surroundings, then it must emit radiation at the same rate as it absorbs it. A perfect black body is therefore an ideal radiator.

The wave theory of radiation could not explain the shape of the black-body radiation curves (Fig 14); in fact the wave theory predicted that infinite amounts of energy would be emitted at short wavelengths (an impossible situation which became known as the 'ultraviolet catastrophe'). In 1900 Max Planck was able to explain the black-body curves by suggesting that the energy was emitted intermittently in packets, called **quanta**, of energy. A radiating body may emit an integral number of these packets of energy, say one, two, three, etc., but cannot emit any fractional amount. The amount of energy, E, carried by each quantum depends on the frequency, f, of the oscillations that are causing the radiation:

$$E \propto f$$

or

$$E = hf$$

where h is the Planck constant, which has a value of 6.6×10^{-34} J s. The packets or quanta of electromagnetic energy are known as **photons**.

> *Example*
>
> *A sodium vapour light emits 30 W of light energy. How many photons are emitted per second? (The wavelength of sodium light is 5.88×10^{-7} m.)*
>
> *Answer*
>
> The frequency of sodium light is given by
>
> $$f = \frac{c}{\lambda} = \frac{3 \times 10^8}{5.88 \times 10^{-7}} = 5.1 \times 10^{14} \text{ Hz}$$
>
> The energy of one photon of light is
>
> $$E = hf = 6.6 \times 10^{-34} \times 5.1 \times 10^{14} = 3.4 \times 10^{-19} \text{ J}$$
>
> Since the sodium light transfers 30 joules per second, the number of photons per second is
>
> $$\frac{30}{3.4 \times 10^{-19}} = 8.8 \times 10^{19} \text{ s}^{-1}$$

Fundamental forces

What forces hold the various particles together in atoms or nuclei? Current theories suggest that there are only four types of interaction between particles.

- **Gravity**. The gravitational force has an infinite range and acts on all particles. Although on the scale of the Universe it is the most important of all the fundamental interactions, on an atomic scale it has negligible influence. This is because gravity is the weakest of all the fundamental forces.

- **Electromagnetic force**. The electromagnetic force acts between all charged particles. Because the electromagnetic force holds atoms and molecules together, it is responsible for almost everything that happens to us. Forces such as friction, buoyancy and contact forces are all electromagnetic in origin.

- **Weak interaction**. The weak interaction acts between all particles but over a very short range, about 10^{-18} m. Over this range it is much stronger than gravitation, 10^{33} times as strong in fact. The weak force is responsible for radioactive decay.

- **Strong interaction**. The strong interaction, or strong nuclear force, holds nuclei together. It acts between **hadrons** (see later), such as neutrons and protons. It is a short-range force, acting over the nuclear distance scale of around 10^{-15} m. The strong interaction is not felt by **leptons** (see later), such as electrons.

Exchange particles

The Japanese physicist Hideki Yukawa suggested that when two particles A and B exert a force on each other a 'virtual particle' is created. This virtual particle can travel between particles A and B and affect their motion.

> An **exchange particle** is a virtual particle, which may exist for only a short time, and is the mediator of a force.

The idea of a force being carried by an exchange particle can be pictured by considering two people on skates (Fig 15). If one of them throws a ball to the other one, both the skaters' motions will be affected; in fact they will be repelled from each other. We have to stretch the analogy a bit to understand attraction, but if you imagine a boomerang being thrown rather than a ball, then the two skaters will be drawn together.

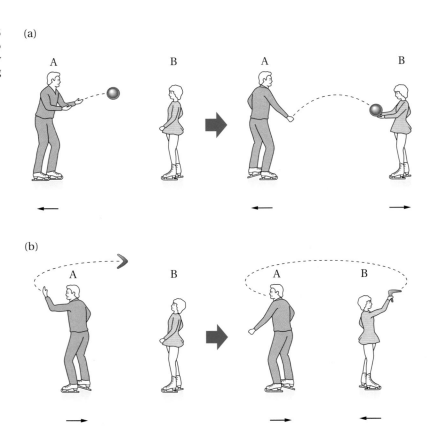

Fig 15
Interactions that can occur when two ice skaters exchange (a) a heavy object, (b) a boomerang

The exchange particles which are transferred between fundamental particles are known as **gauge bosons**. Each fundamental force has its own boson or bosons.

The electromagnetic force
The electromagnetic force is carried between charged particles by the **photon**, γ. When two charged particles exert a force on each other, a photon is exchanged between them. The photon is a massless, chargeless particle. It is its own antiparticle. We use Feynman diagrams to represent what happens (Fig 16).

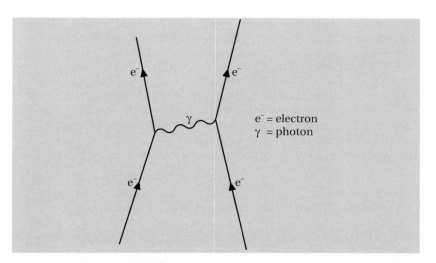

Fig 16
Feynman diagram showing two electrons feeling the electric force as a result of exchange of a photon

The Feynman diagram represents the interaction between the particles. The angles of the particle paths are not significant.

The strong interaction

It was the strong interaction, acting between nucleons, that Yukawa was working on when he proposed the idea of exchange particles. He suggested that these exchange particles could be travelling at close to the speed of light across the nucleus. But where does the energy to create these particles come from? According to Heisenberg's Uncertainty Principle, particles of energy ΔE can be created for a time Δt, provided that the product $\Delta E \times \Delta t$ does not exceed a certain value known as h, the Planck constant, which is a very small number, 6.626×10^{-34} J s. The Uncertainty Principle allows particles to appear for a short time before being annihilated again, provided that $\Delta E \times \Delta t < h$.

An exchange particle moving at close to the speed of light has to exist for about 10^{-23} s if it is to have time to travel across the nucleus. This enabled Yukawa to predict a maximum mass for the particle. The particle was discovered in 1947 and was known as a **pi meson** or **pion** (Fig 17).

The Uncertainty Principle is one of the key ideas of quantum physics. Although it seems at odds with common sense, the theory is in excellent agreement with experimental results. It means that what we used to imagine as a vacuum is not nothing at all, but rather a constant creation and destruction of virtual particle–antiparticle pairs.

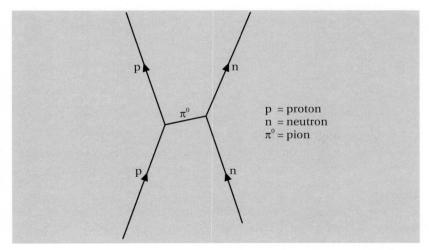

Fig 17
Feynman diagram showing a pion being exchanged between a proton and a neutron

At a deeper level the strong interaction is mediated by gauge bosons called **gluons** that pass between **quarks** (see later). The pion is simply a vehicle carrying gluons between hadrons. There are eight different gluons, none of

Particles are denoted using a symbol, often a Greek letter such as π, Δ or Σ. A superscript is used to denote charge, e.g. π^+ carries a single positive charge, π^- carries a single negative charge and π^0 is uncharged.

which has ever been detected as an individual particle, though scattering experiments have given a strong indication that the theory is correct.

The weak interaction

The weak interaction has a very short range. This suggests that its gauge bosons are relatively massive, since a large mass, i.e. a high energy, would mean a short lifetime and therefore the exchange particles could only travel a small distance. The weak interaction has three gauge bosons, known as the intermediate vector bosons W^+, W^- and Z. These bosons were discovered in 1983 at CERN in Geneva.

The weak interaction acts on leptons and on hadrons. In fact it is the only force, other than gravity, which acts on neutrinos. This explains the fact that neutrinos are so reluctant to interact with anything.

Fig 18
Examples of Feynman diagrams. All these reactions have been observed. They are all due to the weak interaction and are mediated by a W boson. Charge is conserved in all cases

beta-plus (positron) decay: a proton decays into a neutron, emitting an electron neutrino and a positron. The decay occurs via the weak interaction and is mediated by a W^+ boson:
$p \rightarrow n + \nu_e + e^+$

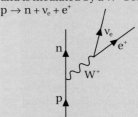

electron capture: an atomic electron can be absorbed by a proton in the nucleus in a process called electron capture. The decay occurs via the weak interaction and is mediated by a W^+ boson: $p + e^- \rightarrow n + \nu_e$

neutrino–neutron collisions: there is a small probability that a neutron can absorb an electron neutrino, emitting a proton and an electron. The reaction occurs via the weak interaction and is mediated by a W^+ boson:
$n + \nu_e \rightarrow p + e^-$

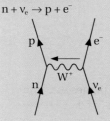

antineutrino–proton collisions: there is a small probability that a proton can absorb an electron antineutrino, emitting a neutron and a positron. The reaction occurs via the weak interaction and is mediated by a W^+ boson:
$p + \bar{\nu}_e \rightarrow n + e^+$

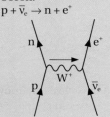

electron–proton collisions: an electron can collide with a proton, emitting a neutron and an electron neutrino. The reaction occurs via the weak interaction and is mediated by a W^- boson:
$p + e^- \rightarrow n + \nu_e$

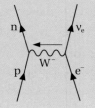

Gravity

The gauge boson which carries the gravitational force is named the **graviton**. It is predicted to have zero rest mass and zero charge. It has never been detected.

10.1.4 Classification of particles

Current theories suggest that there are only three families of fundamental particles: the **leptons**, the **quarks** and the **gauge bosons**.

Leptons

Leptons are fundamental particles; they have no internal structure and are not affected by the strong interaction. There are 12 different particles in the lepton family. The most familiar lepton is the electron. Two other particles, the **muon** and the **tau**, are similar to the electron but more massive. Each has an associated neutrino. All these particles have an antiparticle of opposite charge (see table 4).

The term 'lepton' comes from the Greek word *leptos*, meaning 'small'.

Lepton	Symbol	Charge (in terms of proton charge)	Mass (in terms of electron mass)
electron	e^-	-1	1
electron neutrino	ν_e	0	≈0??
muon	μ^-	-1	207
muon neutrino	ν_μ	0	≈0??
tau	τ^-	-1	3500
tau neutrino	ν_τ	0	≈0??

Table 4
Leptons: each of these leptons has an antiparticle with the same mass, but opposite charge

The muon

The **muon**, μ^-, was discovered in cosmic-ray studies in 1937. The muon is a close relative of the electron. It carries the same charge as the electron but its mass is about 207 times greater.

In 1962 it was shown that the neutrinos that accompany muons are not the same as electron neutrinos. The muon neutrino ν_μ and its antiparticle $\bar{\nu}_\mu$ are also fundamental particles which carry no charge.

In 1978 yet another member of the lepton family was discovered. The **tau minus** particle, τ^-, was observed by a team working on electron–positron collisions at Stanford in the USA. The tau particle has the same charge as the electron and the muon but is around 3500 times the mass of the electron. It is assumed that this new, heavier version of the electron has its own type of neutrino and antineutrino, the ν_τ and the $\bar{\nu}_\tau$. These have not yet been observed (summer 1999).

Leptons: a summary

- The **leptons**, and their antiparticles, **antileptons** (Table 4), are believed to be fundamental particles.
- There are three charged leptons: the **electron**, the **muon** and the **tau particle**.
- Each of these charged leptons has an associated **neutrino**.
- Leptons are not affected by the strong interaction.

Hadrons

In addition to the leptons, the particles now known as pi mesons, kaons and delta mesons had been discovered by the 1960s. All of these have masses much larger than the leptons. This group of particles were referred to as **hadrons**.

'Hadron' is from the Greek word *hadros*, meaning 'bulky'.

By the late 1960s a large number of hadrons had been discovered. Some carried charge and others did not. All except the proton were found to be unstable. Like radioactive atoms, after a certain time they decay into something else. In fact, all hadrons eventually decay into a proton. Even the neutron is unstable when it is free of the nucleus. The neutron decays

The proton is now also considered to be unstable, though its half-life may be of the order of 10^{32} years, much longer than the life of the universe so far.

> **E** You need to know this equation.

with a half-life of around 11 minutes via the reaction

$$n \rightarrow p + e^- + \bar{\nu}_e$$

The hadrons are themselves divided into two groups, the **baryons** and the **mesons**. The baryons, which were orginally thought to be the heavier group, include the proton and the neutron and their antiparticles. The mesons include a large number of particles originally found in cosmic rays, which are now commonly created in collisions inside particle accelerators. The pi meson, or pion, was one of the first to be discovered. It exists in three different forms, positively charged, negatively charged and uncharged. These particles are written π^+, π^- and π^0. Many other mesons have since been discovered, all of them unstable, usually with very short lifetimes. One of the most puzzling at first was the K meson, or kaon, which had a much longer lifetime than other, apparently similar mesons.

10.1.5 Quarks and antiquarks

Hadron reactions and conservation laws

A large number of hadron reactions have been studied. It became apparent that some reactions which appeared to be possible never took place. It seemed as if some reactions were forbidden. We now know that there are physical quantities which cannot change in particle reactions. These *conserved* quantities govern which reactions can take place.

Conservation of charge

Charge is a familiar idea. We know that charged bodies can exert a force on each other and that there are two 'types' of charge which we call positive and negative. Charge is extremely important in particle physics. Many hadrons, and leptons, carry a charge. It is usual to define the charge of the proton to be $+1$. Then the charge of the electron is -1, that of the positron $+1$ and that of the neutron 0.

One of the rules that governs the interactions between particles is the conservation of charge (Table 5). No reactions that contravene this rule have ever been observed.

> **D** The **conservation of charge** means that the total charge after a reaction is the same as the total charge before the reaction.

Table 5 Examples of particle interactions and conservation of charge Q

	Before		After	
reaction	$p + p$	$\rightarrow$	$p + p + \pi^- + \pi^+$	
charge Q	$1 + 1 = 2$		$1 + 1 + (-1) + 1 = 2$	allowed
reaction	$p + \pi^-$	$\rightarrow$	$p + \pi^+$	
charge Q	$1 + (-1) = 0$		$1 + 1 = 2$	not allowed

'Baryon' is from the Greek word *barus*, meaning 'heavy'. This is because baryons were thought to be heavier than mesons. This is not always true.

Conservation of baryon number

There are some reactions allowed by charge conservation which have never been observed. This is because there are other conservation laws which place restrictions on which reactions can take place. One of these is the conservation of **baryon number**, B.

All particles which are not hadrons, i.e. the leptons and the gauge bosons, have a baryon number of 0.

The hadrons can be divided into three groups: the **mesons**, the **baryons** and the **antibaryons** (Table 6). Mesons have a baryon number of 0, baryons 1 and antibaryons -1. Reactions between any of these hadrons can only occur if the baryon number is conserved. Just like charge, the total baryon number before the interaction has to be the same as the total baryon number afterwards (Table 7).

Mesons B = 0	Baryons B = 1	Antibaryons B = −1
pi mesons (pions) π^+, π^-, π^0 K mesons (kaons) K^+, K^-, K^0	proton p and neutron n sigma particles Σ^+, Σ^-, Σ^0	antiproton $\bar{p}$ antineutron $\bar{n}$

Table 6
Hadrons (B = baryon number)

	Before		After	
reaction B	p + p 1 + 1 = 2	→	p + p + n 1 + 1 + 1 = 3	not allowed
reaction B	p + p 1 + 1 = 2	→	p + p + π^0 1 + 1 + 0 = 2	allowed

Table 7
Particle interactions and conservation of baryon number B

Conservation of strangeness

The rules of conservation of charge and conservation of baryon number do not fully explain why some reactions are never observed.

K mesons, or **kaons**, caused particular problems to particle physicists. Kaons appear as the decay products of some neutral particles, but they always seemed to turn up in pairs. Kaons didn't appear individually, although charge or baryon number conservation would not prevent this. They also had an unusually long lifetime, 10^{-10} s, compared with other, apparently similar, hadrons, which had typical lifetimes of the order of 10^{-23} s.

There is another property that has to be conserved in hadron reactions. This property is called **strangeness**. All hadrons are given a strangeness number, S, of either $+1$, 0, -1 or -2. Strangeness has to be conserved in any reaction that takes place via the strong interaction (Tables 8 and 9).

Unlike charge conservation and conservation of baryon number, there are some interactions where strangeness is not conserved. These are reactions that take place via the weak interaction.

S = −2	S = −1	S = 0	S = +1
Ξ^- (xi minus) Ξ^0 (xi zero)	Λ (lambda) K^- (K minus) Σ^+, Σ^-, Σ^0 (sigma particles)	p (proton) n (neutron) π^+, π^-, π^0 (pions)	K^+ and K^0 (kaons)

Table 8
Strangeness S for some hadrons

	Before		After	
reaction S	p + π^- 0 + 0 = 0	→	K^0 + Λ^0 1 + (−1) = 0	allowed
reaction S	p + π^- 0 + 0 = 0	→	K^- + Σ^+ (−1) + (−1) = −2	not allowed

Table 9
Particle interactions and conservation of strangeness S

Allowed reactions

All hadrons have fixed values for charge Q, baryon number B and strangeness S. Reactions between hadrons can only take place if the reaction conserves these numbers.

When strange particles, i.e. those with non-zero values of strangeness, decay, they do so by the weak interaction and strangeness is *not* conserved. In these reactions strangeness changes by ±1.

> Check that Q, B and S are the same on both sides of the equation.

Example

Which of these reactions is not allowed?

(i) $\pi^+ + p \rightarrow K^+ + n$

(ii) $K^+ + \bar{p} \rightarrow \pi^0$

(iii) $p + p \rightarrow p + p + \pi^0$

Answer

(i) Cannot occur, as it violates charge conservation and strangeness conservation.

(ii) Cannot occur, as it violates baryon number and strangeness conservation.

(iii) Can occur.

Quarks

In the 1960s a scattering experiment carried out at Stanford in the USA revealed the structure inside hadrons.

The Stanford Linear Accelerator Center (SLAC) in California could accelerate electrons to an energy of around 6 GeV, high enough to probe the structure of nucleons, i.e. to look inside protons and neutrons. The SLAC experiment found that a significant proportion of high-energy electrons were scattered through a large angle. This indicated that the neutrons and protons are not particles of uniform density, but have point-like charged particles within them (Fig 19).

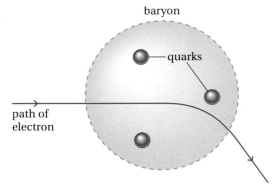

Fig 19
Electron scattering by quarks within a baryon

> The term 'quark' was given to these particles by Murray Gell-Mann. He took the term from 'Three quarks for Muster Mark', a quotation from *Finnegans Wake* by James Joyce.

The SLAC results confirmed a theory put forward by Murray Gell-Mann and George Zweig a few years earlier. Gell-Mann had grouped the hadrons together in families. The patterns could be explained by supposing that all hadrons were composed of smaller constituents, named **quarks**. The SLAC experiments confirmed that hadrons are not fundamental particles but are composed of combinations of different types of quark.

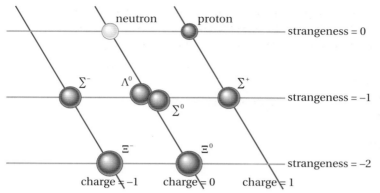

Fig 20
A family of baryons

These eight baryons all have a baryon number of 1. There is a similar grid for their antiparticles (baryon number −1). The Λ^0 and Σ^0 differ only in their energy.

Gell-Mann called the patterns (Fig 20) the 'eightfold way', a term from Buddhism.

Gell-Mann suggested that there were three different quarks, labelled 'up' (u), 'down' (d) and 'strange' (s). Each of these quarks has a specific mass and values for charge, baryon number and strangeness. Each quark has a corresponding antiquark of exactly equal mass but opposite values for charge, baryon number and strangeness (Table 10).

Using the quark model it is possible to describe all hadrons in terms of combinations of quarks and antiquarks (Fig 21). **Baryons** are combinations of three quarks, not necessarily the same type. **Antibaryons** are combinations of three antiquarks. **Mesons** are composed of a quark and an antiquark, again not necessarily the same type.

Table 10
Properties of quarks and antiquarks

Quark	Baryon number B	Charge Q	Strangeness S	Anti-quark	Baryon number B	Charge Q	Strangeness S	Mass (GeV/c^2)
up, u	$\frac{1}{3}$	$\frac{2}{3}$	0	$\bar{u}$	$-\frac{1}{3}$	$-\frac{2}{3}$	0	0.005
down, d	$\frac{1}{3}$	$-\frac{1}{3}$	0	$\bar{d}$	$-\frac{1}{3}$	$+\frac{1}{3}$	0	0.01
strange, s	$\frac{1}{3}$	$-\frac{1}{3}$	−1	$\bar{s}$	$-\frac{1}{3}$	$+\frac{1}{3}$	+1	0.2

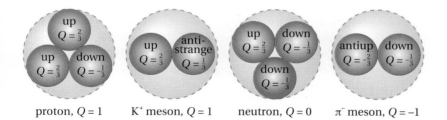

Fig 21
Neutron, K^+ meson, proton and π^- meson, showing quark structure

The properties of each hadron can be explained in terms of the quarks that it is made from. The total charge on the hadron is the sum of the quark charges. The same can be said for the total baryon number and strangeness.

Table 11
Quark structure of a proton:
$p = u + u + d$

	Up, u	Up, u	Down, d	Proton, p
charge Q	$\frac{2}{3}$	$\frac{2}{3}$	$-\frac{1}{3}$	1
baryon number B	$\frac{1}{3}$	$\frac{1}{3}$	$\frac{1}{3}$	1
strangeness S	0	0	0	0

Table 12
Quark structure of an antiproton:
$\bar{p} = \bar{u} + \bar{u} + \bar{d}$

	Antiup, $\bar{u}$	Antiup, $\bar{u}$	Antidown, $\bar{d}$	Antiproton, $\bar{p}$
charge Q	$-\frac{2}{3}$	$-\frac{2}{3}$	$\frac{1}{3}$	-1
baryon number B	$-\frac{1}{3}$	$-\frac{1}{3}$	$-\frac{1}{3}$	-1
strangeness S	0	0	0	0

Table 13
Quark structure of a neutron:
$n = u + d + d$

	Up, u	Down, d	Down, d	Neutron, n
charge Q	$\frac{2}{3}$	$-\frac{1}{3}$	$-\frac{1}{3}$	0
baryon number B	$\frac{1}{3}$	$\frac{1}{3}$	$\frac{1}{3}$	1
strangeness S	0	0	0	0

Table 14
Quark structure of a π^- meson:
$\pi^- = \bar{u} + d$

	Antiup, $\bar{u}$	Down, d	π^-
charge Q	$-\frac{2}{3}$	$-\frac{1}{3}$	-1
baryon number B	$-\frac{1}{3}$	$\frac{1}{3}$	0
strangeness S	0	0	0

Table 15
Quark structure of a K^+ meson:
$K^+ = u + \bar{s}$

	Up, u	Antistrange, $\bar{s}$	K^+
charge Q	$\frac{2}{3}$	$\frac{1}{3}$	1
baryon number B	$\frac{1}{3}$	$-\frac{1}{3}$	0
strangeness S	0	1	1

Table 16
Description of pions and kaons in the simple quark model

π^+	$u\bar{d}$
π^-	$\bar{u}d$
π^0	$u\bar{u}$ or $d\bar{d}$
K^+	$u\bar{s}$
K^-	$\bar{u}s$
K^0	$d\bar{s}$
$\bar{K}^0$	$\bar{d}s$

Only two types of quark, the up and the down, are needed to account for the properties of the neutrons and protons which together make up almost all of everyday, observable matter. The two antiquarks $\bar{u}$ and $\bar{d}$ are needed to explain the existence of the antiproton and the antineutron.

The properties of a meson are the sum of the properties of its constituent quarks. A meson's antiparticle is the opposite combination of quark and antiquark. For example the pi plus meson is formed from the two quarks $u\bar{d}$, whilst its antiparticle, the pi minus, is formed from the opposite combination $\bar{u}d$ (Table 16). The K^- meson is formed from the quarks $\bar{u}s$ (Table 15).

The quark model has been very successful in describing and predicting the properties of hadrons. In beta-minus emission, for example, a neutron decays to a proton, emitting an electron and an antineutrino in the process (Fig 22). The quark theory tells us that inside the neutron a down quark has changed into an up quark, emitting an electron and an antineutrino.

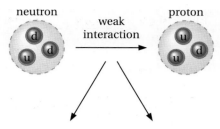

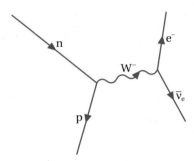

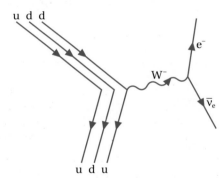

Fig 22
Beta-minus emission

10.2 Electromagnetic radiation and quantum phenomena

Electromagnetic waves

Electromagnetic waves are emitted by the oscillation of charged particles, such as an electron. The oscillation sets up varying electric and magnetic fields which travel through space. The electric and magnetic fields are at right angles to each other and to the direction of travel of the waves (Fig 23). The wave propagates through space as a transverse wave, without the need for any supporting medium.

> A **transverse wave** has oscillations at right angles to its direction of travel, whereas a **longitudinal wave** (like sound) has oscillations that are parallel to the direction of travel.

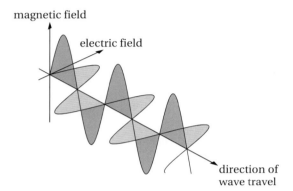

Fig 23 An electromagnetic wave

All electromagnetic waves travel at the same speed in a vacuum, that is, 2.98×10^8 m s^{-1}. However, the properties of the wave and the way that it interacts with matter depend on its wavelength.

> **D** The distance between any two identical points on a wave is the **wavelength**, λ.

This is measured in metres. For a pure sine wave it is the distance between any two adjacent crests or troughs.

> **D** The time taken for a wave to go through one complete oscillation is the **period**, T.
> The number of oscillations per second, f, is the **frequency**.

The frequency is measured in hertz (Hz). Frequency $= \dfrac{1}{\text{period}}$, that is,

$$f = \frac{1}{T}$$

Since speed = distance divided by time, the speed that the wave travels at, c, is given by $c = \dfrac{\lambda}{T}$

or $c = \lambda \times \dfrac{1}{T}$

> **E** You need to know the equation $c = f\lambda$.

So $c = f\lambda$

Light

Light is an electromagnetic wave. Visible light has a wavelength range from about 400 nm (violet) to around 700 nm (red).

Example

Find the frequency range of visible light.

Answer

Since

$c = f \times \lambda$

$f = \dfrac{c}{\lambda}$

If $\lambda = 400 \times 10^{-9}$ m, then

$f = \dfrac{3 \times 10^8}{400 \times 10^{-9}} = 7.5 \times 10^{14}$ Hz

If $\lambda = 700 \times 10^{-9}$ m, then

$f = \dfrac{3 \times 10^8}{700 \times 10^{-9}} = 4.3 \times 10^{14}$ Hz

---

Different wavelengths, or frequencies, of light give us the impression of different colours. Larger-amplitude waves increase the intensity (brightness) of the light.

Like the rest of the electromagnetic spectrum, light can be reflected, refracted and diffracted.

Reflection of light waves

The wave theory can be used to explain the **reflection** of light. Each point on a wavefront can be thought of as being a new source of waves. These new waves, called **secondary wavelets**, later form the new wavefront. The idea of secondary wavelets is known as **Huygens' construction** (Fig 24). Secondary wavelets from the reflecting surface form the new wavefront (Fig 25).

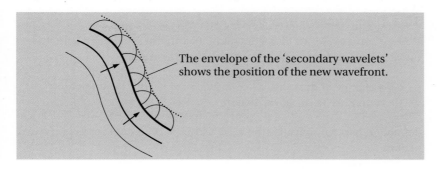

Fig 24
Huygens' construction

Fig 25
Huygens' construction for reflection at a plane surface

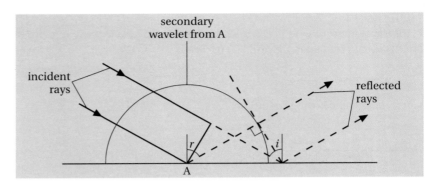

An easier way to represent this is to draw lines, known as **rays** (Fig 26), at right angles to the wavefronts, which show the direction of motion of the wave.

Fig 26
Reflection represented by rays

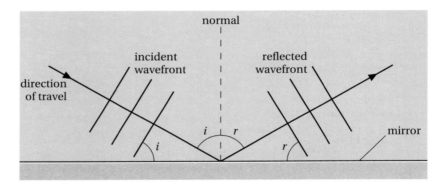

By convention, angles are measured from the **normal**, a line drawn at right angles to the surface (Fig 26). For any reflecting surface, the laws of reflection are as follows:

- The angle of incidence i and the angle of reflection r (both measured from the normal) are equal.
- The incident ray, the reflected ray and the normal all lie in the same plane.

10.2.1 *Refraction at a plane surface*

When a wave moves from one medium into another, such as a light wave moving from air into glass, it usually changes direction.

> *The change of direction when a wave moves from one medium into another is called refraction.*

Refraction happens because the wave travels at different speeds in the two media.

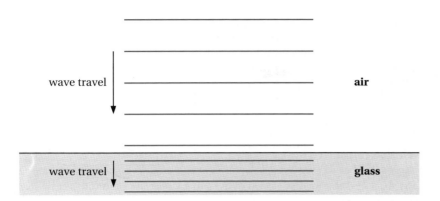

Fig 27
Light waves travelling normally into a pane of glass, showing the change of wavelength. Here there is no change of direction, but the wavelength and speed of the waves are reduced

The speed of light in air is almost 3×10^8 m s^{-1}, but in glass it is much less, about 2×10^8 m s^{-1}. As a light wave travels into a glass window the wavefronts become closer together as the wave slows down (Fig 27). The speed c and the wavelength λ both decrease but the frequency f stays the same.

If the wavefront hits the boundary between the two media at an angle to the normal then the wavefront changes direction. A ray of light refracts towards the normal when it slows down, and away from the normal when it speeds up (Fig 28). The greater the change in speed, the more the ray deviates from its original path.

When light waves pass through glass, the frequency of the waves must stay the same. After all if 5×10^{14} waves per second enter the glass then the same number must come out – if not, where would they go?

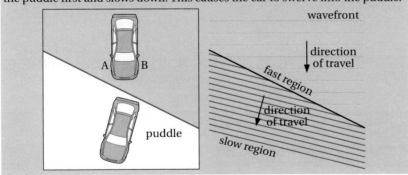

Fig 28
A change of speed

Refractive index

All electromagnetic waves, including light waves, travel at the same speed c in a vacuum. Precise measurements give this value as $c = 299\,792\,458$ m s^{-1}. As the waves pass through matter they slow down to a new speed v.

> **The absolute refractive index n of a material is the ratio of the speed of light in a vacuum to its speed in the material.**

$$n = \frac{\text{speed of light in a vacuum}}{\text{speed of light in the material}} = \frac{c}{v}$$

The refractive index of air is 1.0003 but this is usually taken as being equal to 1. In other words we usually assume that light travels at the same speed in air as in a vacuum.

Refractive index is a ratio and has no units.

Table 17
Refractive indexes

diamond	2.42
glass	1.5 to 2.0
Perspex	1.50
water	1.33
sea water	1.34
ice	1.31

When we compare the speed of light in a medium with the speed in a *vacuum*, the ratio is sometimes referred to as the **absolute refractive index**. Because light always travels faster in a vacuum than in any other material, the value of the absolute refractive index is always greater than 1. Refractive indexes of some common materials are listed in Table 17. The higher the value of the refractive index, the more the light is slowed down and the greater its deflection. Materials with a high value of refractive index are said to be 'optically dense'.

Relative refractive index

When a light wave passes from one material into another, say from water into glass, it is the relative speed of light in each of the materials that determines how much a ray will be deflected. The **relative refractive index** is

$$_1n_2 = \frac{\text{speed of light in medium 1}}{\text{speed of light in medium 2}} = \frac{v_1}{v_2}$$

If we know the absolute refractive indexes for two materials, say n_1 and n_2, it is possible to calculate the relative refractive index for a light wave moving between them.

The relative refractive index for a wave moving from medium 1 to medium 2 is the inverse of the relative refractive index for a wave moving in the opposite direction, from medium 2 to medium 1:

$$_1n_2 = \frac{1}{_2n_1}.$$

$$_1n_2 = \frac{v_1}{v_2} = \frac{c}{v_2} \times \frac{v_1}{c} = n_2 \times \frac{1}{n_1} = \frac{n_2}{n_1}$$

So

$$_1n_2 = \frac{n_2}{n_1}$$

The relative refractive index can have a value of less than one. For example, the relative refractive index for a light wave moving from water to ice is

$$_{\text{water}}n_{\text{ice}} = \frac{n_{\text{ice}}}{n_{\text{water}}} = \frac{1.31}{1.33} = 0.985$$

Snell's law of refraction

The relative refractive index between two materials affects how much a ray of light will deviate at the boundary.

$$_1n_2 = \frac{\sin(\text{angle of incidence})}{\sin(\text{angle of refraction})}$$

$$= \frac{\sin\theta_1}{\sin\theta_2}$$

Another way to put this is that a ray of light will refract *towards* the normal as it enters an optically denser medium, and *away* from the normal as it leaves again.

This is **Snell's law** (Fig 29).

If the light waves are travelling into an optically denser material, say from air into water, they will slow down, and θ_2 will be less than θ_1. The angle is always smaller in the optically denser material.

Fig 29
Snell's law

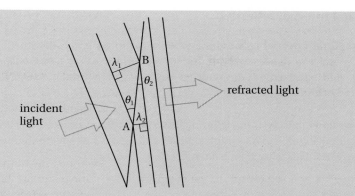

Adjacent wavefronts are separated by one wavelength. As the wave crosses the boundary, its frequency remains constant so the change of speed causes a change of wavelength, from λ_1 to λ_2. By definition, $v = f\lambda$ and $_1n_2 = \frac{v_1}{v_2}$ so

$$_1n_2 = \frac{f\lambda_1}{f\lambda_2}$$

$$= \frac{\lambda_1}{\lambda_2}$$

Using $\lambda_1 = AB \sin \theta_1$ and $\lambda_2 = AB \sin \theta_2$, $\quad _1n_2 = \frac{AB \sin \theta_1}{AB \sin \theta_2} = \frac{\sin \theta_1}{\sin \theta_2}$

It is easy to show that θ_1 is equal to the angle of incidence and θ_2 is equal to the angle of refraction.

> Snell's law applies (a) when a light ray passes from one medium to another, (b) for monochromatic light (single wavelength) and (c) to rays which lie in the same plane.

Example

A ray of light strikes a glass block at an angle of 30° to the normal. Find the angle of refraction in the glass block. (Take the speed of light in glass to be 2.0×10^8 m s^{-1} and the speed of light in air to be 3.0×10^8 m s^{-1}.)

Answer

Refractive index of glass $= \dfrac{3.0 \times 10^8 \text{ m s}^{-1}}{2.0 \times 10^8 \text{ m s}^{-1}} = 1.5$

$_1n_2 = \dfrac{\sin \theta_1}{\sin \theta_2}$

$\theta_1 = 30°$

So

$\sin \theta_2 = \dfrac{\sin 30°}{1.5} = 0.33$

Therefore

$\theta_2 = 19.5°$

Total internal reflection

> **Total internal reflection** is when a ray of light, leaving an optically dense material and travelling into a less dense one, is not refracted out of the dense material but is totally reflected back inside.

The light could be moving from glass into air, for example. If the angle of incidence in the glass is high enough, the light is no longer refracted out of the glass. This is total internal reflection.

Fig 30 Refraction and reflection

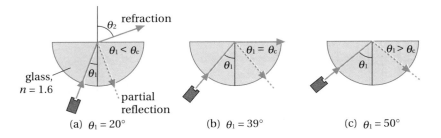

(a) $\theta_1 = 20°$ (b) $\theta_1 = 39°$ (c) $\theta_1 = 50°$

Total internal reflection occurs when (a) the light ray is incident inside the optically denser medium *and* (b) the angle of incidence is greater than the critical angle.

We can use Snell's law to calculate the angle of refraction in each of the cases shown in Fig 30.

If the refractive index from air to glass is 1.6, then the refractive index from glass to air is

$$\frac{1}{1.6} = 0.625$$

(a) $\sin \theta_2 = \dfrac{\sin \theta_1}{n} = \dfrac{\sin 20°}{0.625} = 0.547$

so the angle of refraction is 33.2°.

(b) $\sin \theta_2 = \dfrac{\sin \theta_1}{n} = \dfrac{\sin 39°}{0.625} = 1.00$

so the angle of refraction is 90°. This means that the refracted ray is deviated so much that it just grazes the surface of the block.

(c) $\sin \theta_2 = \dfrac{\sin \theta_1}{n} = \dfrac{\sin 50°}{0.625} = 1.23$

It is impossible for the sine of an angle to be greater than 1. Snell's law is not applicable here because the ray is no longer refracted, but is totally reflected inside the block.

The change from refraction to reflection is not sudden; some light is always reflected inside the block. As the angle of incidence increases, more and more of the light is reflected. At one particular angle of incidence, the **critical angle**, the refracted ray disappears (Fig 30(b)). At this angle the ray is trying to travel along the boundary between the two materials. If the angle of incidence is greater than the critical angle then *all* the incident light is reflected and none of it escapes.

The value of the critical angle depends on the refractive indexes of the two media. At the critical angle θ_c, the angle of refraction is 90°, so

$$_1n_2 = \frac{\sin\theta_c}{\sin 90°} = \sin\theta_c$$

Also

$$_1n_2 = \frac{n_1}{n_2}$$

so

$$\sin\theta_c = \frac{n_2}{n_1}$$

Where the rays of light are moving from a medium of refractive index n into air, which has a refractive index of 1, then

$$\sin\theta_c = \frac{1}{n}$$

Example

An optical fibre is made from glass of refractive index 1.5. A ray of light strikes the end of the fibre at an angle of 75° to the normal. Sketch the situation, showing the path of the ray through the fibre.

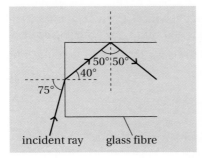

Answer

Since the angle of incidence is 75°, and the refractive index is 1.5, the angle of refraction is given by

$$\sin\theta_2 = \frac{\sin 75°}{1.5}$$

So

$$\theta_2 = 40°$$

The ray will strike the wall of the fibre at an angle of 50°. This is above the critical angle for glass (since $\sin\theta_c = 1/n = 1/1.5 = 0.67$, $\theta_c = 42°$).

This means that the ray is totally reflected inside the glass. The reflected ray obeys the rule of reflection.

A prism can be used to reflect light (Fig 31). The reflection is better than with a mirror, where silvering causes multiple reflections.

Fig 31
A prism used in two ways to reflect light

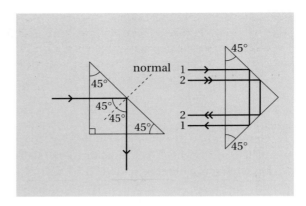

Totally reflecting triangular prisms are used in preference to mirrors in binoculars to invert the image. It is total internal reflection which makes precious stones sparkle; the refractive index of diamond is very high, around 2.4. Total internal reflection is also responsible for the mirage which makes a tarmac road appear shiny on a hot day.

Fibre optics

The most important application of total internal reflection is in **optical fibres**. Optical fibres carry cable TV and telephone communications and form the backbone of computer networks.

The number of digital (on–off) pulses that can be transmitted per second along a fibre is known as its **bandwidth**.

In its simplest form an optical fibre is a thin strand of pure glass. Light is refracted into one end and strikes the internal wall of the fibre at an angle of incidence greater than the critical angle. Total internal reflection occurs and the ray of light is confined within the fibre (Fig 32).

If the refractive index of the glass fibre is 1.60, then the critical angle is given by

$$\sin \theta_c = \frac{1}{n} = \frac{1}{1.60} = 0.625$$

This gives a critical angle of $\sin^{-1} 0.625$ or 39°. Any rays of light that strike the inner surface of the glass at angles greater than this will be reflected inside the fibre.

Fig 32
General path of light through an optical fibre

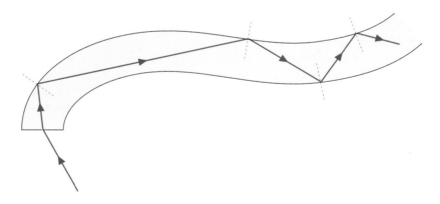

Optical fibres offer much greater bandwidth than conventional copper cables.

Optical fibres are used to carry information in the form of digital pulses over long distances. It is important that the light losses are as low as possible. This is achieved by using glass with a very low level of impurities, since impurity atoms can scatter the light so that it strikes the boundary at less than the critical angle and is refracted out of the fibre.

Scratches on the surface of the fibre are another potential source of signal loss (Fig 33). One way to avoid this is to use an outer layer of different glass, known as **cladding**, to protect the inner, or core, fibre. The cladding has to be made from a glass of lower refractive index than the core, otherwise total internal reflection could not take place.

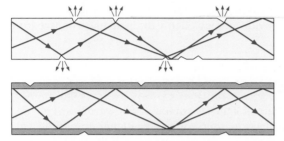

Surface scratches lead to light leakage in a single fibre. Scratches in the outer cladding do not matter because it does not carry a light signal.

Fig 33
Light leakage in an optical fibre due to scratches

For an optical fibre with a core of refractive index $n_1 = 1.60$ and a cladding of $n_2 = 1.50$, the critical angle is given by

$$\sin \theta_c = \frac{n_2}{n_1} = \frac{1.5}{1.6} = 0.9375$$

So

$$\theta_c = 69.6°$$

This is a much larger critical angle than for a single glass fibre in air and so fewer rays will be reflected inside the fibre.

Optical fibres are also used in medicine to enable doctors to look into the body without the need for invasive surgery. In **gastroscopy** a bunch of optical fibres is used to carry light down the oesophagus into the stomach. A second bunch of optical fibres is used to carry the image information back to a video camera. Similar devices, collectively known as **endoscopes**, are used to look inside joints (**arthroscopy**) and other parts of the body. In some cases the fibres can carry intense laser light, which is used instead of a scalpel as a tool in surgery.

10.2.2 *The photoelectric effect*

The **photoelectric effect** was discovered towards the end of the 19th century. Experiments showed that electrons could be emitted from the surface of a metal by illuminating the metal with light. However, some of the experimental results seemed completely at odds with the wave theory of light.

- The electrons are only emitted from the surface of the metal if the light is above a certain frequency. For example, if zinc is illuminated with visible light, no electrons are emitted. It is only when ultraviolet light is used that there is any effect. Every metal has its own particular light frequency, known as the **threshold frequency**, below which there is no photoemission.

- The electrons are emitted with a range of different kinetic energies from zero up to a maximum value. The maximum <u>kinetic energy depends on the frequency of the light,</u> not the intensity. A faint ultraviolet glow would cause the emission of more energetic electrons than an intense red laser beam.
- If the light is above the threshold frequency, then the number of electrons emitted per second is proportional to the intensity of the light.

<u>Electrons are held by electrostatic forces</u> onto the surface of the metal. The light has to provide enough energy to rip an electron free from the metal surface.

> *The energy needed to remove an electron from the surface of a metal is called the **work function**, denoted by ϕ.*

The wave theory of light says that if a light wave hasn't got enough energy to release an electron, then you need a higher-amplitude wave, i.e. a brighter light wave. But this doesn't work. If the light is below the threshold frequency, it doesn't matter how intense it is, there will be no photoemission.

Einstein explained the photoelectric effect by using the idea of photons. He realised that light is absorbed in discrete packets of energy, called photons. When a photon strikes a metal surface, it is absorbed either totally or not at all. So when a photon strikes the surface of a metal and collides with an electron, it will only dislodge an electron if its energy, E, is larger than the work function.

Photoemission only occurs when

$$E > \phi$$

or, since

$$E = hf$$

when

$$hf > \phi$$

When photoemission *just* occurs, the threshold frequency f_0 is given by

$$f_0 = \frac{\phi}{h}$$

Above the threshold frequency, the photon carries more than enough energy to release an electron. The excess energy goes into the kinetic energy of the emitted electron (Fig 34). Einstein's photoelectric equation expresses this in terms of the conservation of energy:

energy of incident photon = energy needed to remove the electron (work function) + kinetic energy of the emitted electron

$$hf = \phi + E_k$$

<u>E_k is the maximum kinetic energy of the electron;</u> if the electron has been emitted from deeper within the metal surface it may have a lower value of kinetic energy.

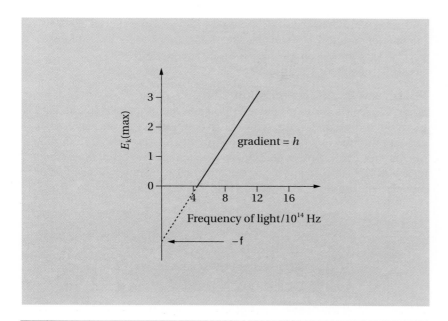

Fig 34
Maximum kinetic energy of emitted electron versus frequency of incident light:
$E_k(\text{max}) = hf - \phi$

This graphs shows exactly what Einstein predicted: that the maximum energy of the released electrons is proportional to the frequency of the light. The constant of proportionality (the gradient of the graph) is equal to h.

Example

Light of wavelength 330 nm is incident on a metal which has a threshold frequency of 5×10^{14} Hz. Find the maximum kinetic energy of the emitted electrons.

Answer

The energy of the incident photons is

$$E = hf = \frac{hc}{\lambda} = 6 \times 10^{-19} \text{ J}$$

The work function of the metal is

$$hf_0 = 6.6 \times 10^{-34} \times 5 \times 10^{14} = 3.3 \times 10^{-19} \text{ J}$$

Using Einstein's photoelectric equation

$$hf = \phi + E_k$$

$$E_k = hf - \phi = 6 \times 10^{-19} - 3.3 \times 10^{-19} = 2.7 \times 10^{-19} \text{ J or 1.7 eV}$$

10.2.3 Collisions of electrons with atoms

The electronvolt

The joule is rather a large unit of energy to use in atomic and nuclear physics. In practice physicists use the electronvolt as a convenient unit of energy.

One joule of energy is transferred when a coulomb of charge passes through a potential difference of one volt. An electronvolt is the energy for one electron passing through 1 volt. The electronvolt is therefore a much smaller unit of energy.

> **An electronvolt (eV) is the amount of energy gained by an electron as it accelerates through a potential difference of 1 volt. 1 eV = 1.6×10^{-19} J.**

Cathode ray tubes such as the one used by Thomson achieved electron energies of around 1 keV (one kiloelectronvolt, or 1×10^3 eV). Modern particle accelerators are pushing towards energies of 1 TeV (teraelectronvolt, or 1×10^{12} eV).

Energy levels, photon emission

When an electric current is passed through a vapour of an element, light is given off. If the light is observed through a diffraction grating, each element is found to have its own set of bright emission lines (Fig 35 shows the lines for hydrogen). For any specific element the spectral lines are always at the same frequency. To explain this we need to look again at our model of the atom.

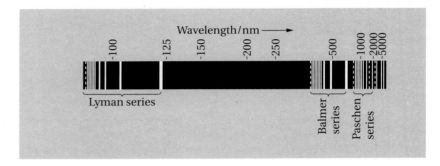

Fig 35
Line series in the hydrogen spectrum

The Balmer series is a set of coloured lines in the visible part of the spectrum. The Lyman and Paschen series are in the ultraviolet and infrared regions, respectively.

Rutherford's model of the hydrogen atom has one electron orbiting a very small, dense, positively charged nucleus. There is a problem with this model. All charged particles emit radiation when they accelerate. The orbiting electron is accelerating towards the centre of its orbit as it constantly changes direction. According to the laws of classical physics the electron should be radiating energy all the time. As it radiates it should lose energy, eventually spiralling down towards the nucleus. This is rather like an artificial satellite that has dropped into a low orbit around the Earth. As the satellite passes through the upper atmosphere it loses energy and so it will inevitably drop further and spiral down towards the Earth's surface.

Niels Bohr suggested that the electron could travel in certain allowed orbits (Fig 36) without losing energy. He called these allowed orbits 'stationary states'. When the electron is in an allowed orbit it does not radiate, but stays at a constant energy.

Bohr said that an electron in an atom can only emit or absorb energy as it moves from one allowed orbit to another (Fig 37). This idea helped to explain the existence of line spectra. Light is emitted from atoms when electrons lose energy, but in Bohr's atom electrons can only lose energy in specified amounts as they tumble down the energy levels. An electron can only move from one allowed state to another by gaining or losing exactly the right amount of energy. That is why only certain frequencies of light appear in the line spectrum.

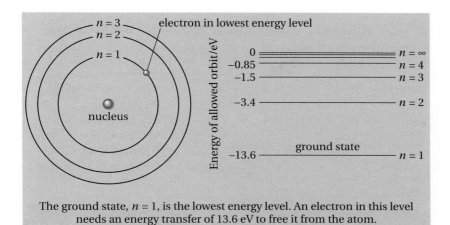

Fig 36
Allowed orbits and energy levels in Bohr's hydrogen atom

The energy levels are negative because the electron is in a bound state – it is tied to the atom. The energy value of each allowed orbit tells you how much energy is needed to free the electron from the atom. By this convention higher energy levels are less negative; an electron with zero energy is just free of the atom.

The ground state, $n = 1$, is the lowest energy level. An electron in this level needs an energy transfer of 13.6 eV to free it from the atom.

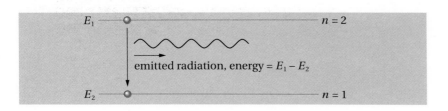

Fig 37
Electron transition

Each time an electron falls to a lower energy level it loses energy. This energy is radiated as a photon of frequency f. So the energy of the emitted photon is

$$E_1 - E_2 = \Delta E = hf$$

Bohr's model of the atom was successful in explaining the existence of line spectra. Although the energy level diagrams are more complicated for atoms other than hydrogen, the same principles apply.

Ionisation, excitation

When the electrons are in their lowest-energy orbits, an atom is said to be in its **ground state**. The lowest allowed orbit for a hydrogen atom has an energy of -13.6 eV. When a hydrogen atom is in the ground state, its electron cannot lose any more energy. The ground state is the preferred state for an atom, but electrons can move to higher energy levels if they absorb the correct amount of energy. This process could be caused by the absorption of a photon of radiation of the right wavelength or by a collision with another electron (Fig 38).

> **Excitation** is when an electron moves to a higher energy level.

> **Ionisation** is when an electron gains so much energy that its total energy becomes positive. This means that it becomes free of the atom.

Fig 38
Excitation by (a) absorption of a photon and (b) collision of an electron; (c) shows ionisation

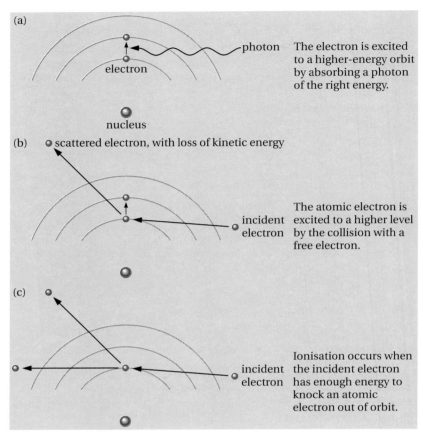

The fluorescent lamp
When an electric current is passed through a fluorescent lamp, electrons collide with atoms of mercury vapour. If the electron has sufficient energy, greater than 6.7 eV, the collision will excite an electron in the mercury atom to a higher energy level. As the atomic electron returns to its original state it emits a photon of ultraviolet light. This ultraviolet light is converted to visible light by the phosphors on the inside of the glass tube.

> *Example*
> *A fast-moving electron with a kinetic energy of 150 eV collides with an electron in a hydrogen atom. Explain what is likely to happen.*
>
> *Answer*
> Assuming the hydrogen atom is in its ground state, its electron needs only 13.6 eV to free it from the atom. The incident electron will transfer some of its kinetic energy to the atomic electron. The collision will cause ionisation and the ejected electron will gain kinetic energy.

10.2.4 Wave–particle duality

Photons or waves?
The wave theory successfully explains the way that light is reflected and refracted and is important in interpreting the phenomena of diffraction and interference. However, the wave theory cannot describe the

photoelectric effect or black-body radiation curves. These have been explained by thinking of light as a stream of massless particles called photons, which carry energy. This is an example of what is referred to as **wave–particle duality**. There is no real contradiction here. The wave and particle theories are complementary to each other. The wave theory gives us an excellent way of picturing what happens as light passes from one place to another, whilst the particle theory is useful in describing how light interacts with electrons (Fig 39).

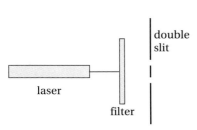

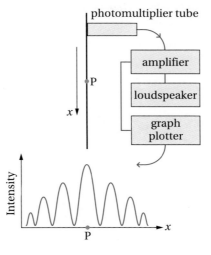

Fig 39
Wave–particle duality experiment

A photomultiplier tube uses the photoelectric effect to detect light photons. It therefore relies on the particle nature of light, yet it detects an interference pattern which can only be caused by waves. If the filter is dense enough, the loudspeaker lets you hear the photons arriving, one by one.

Particles or waves?

It isn't only light that shows aspects of wave and particle behaviour. Electrons, which we have so far treated as point particles, can be made to diffract. Louis de Broglie suggested, in 1924, that 'all material particles should have a wave nature'. He predicted that a particle of momentum p should have a wavelength of λ given by

$$\lambda = \frac{h}{p} \quad \text{or} \quad \lambda = \frac{h}{mv}$$

where h is the Planck constant.

> You should be able to give examples of both particles and electromagnetic waves which support both the wave and the particle nature of each.

This idea can be tested by trying to diffract electrons through a suitable aperture. Remember that diffraction effects become noticeable when the size of the aperture is of the same magnitude as the wavelength of the waves. An electron accelerated across an electric potential of 5 kV will reach a velocity of around 4.2×10^7 m s^{-1}. This gives it a momentum of 3.8×10^{-23} kg m s^{-1}. According to de Broglie's equation, the electron has a wavelength of

$$\lambda = \frac{h}{p} = \frac{6.6 \times 10^{-34}}{3.8 \times 10^{-23}} = 1.7 \times 10^{-11} \text{ m}$$

This is about the size of the gaps between layers of atoms.

> The electron microscope helps us to see much finer detail than is possible using a light microscope. This is because it is diffraction that limits our ability to see fine detail. Electrons diffract very little because they have a very short wavelength.

In 1928, four years after de Broglie put forward his theory, George Thomson (the son of J.J. Thomson) produced an electron diffraction pattern by firing high-speed electrons at a gold foil. The emerging electron beam showed the same variation in intensity as light that had passed through a diffraction grating.

Today electron microscopes, which rely on the wave nature of electrons, are in common use. There are also microscopes which use protons and even ions. Since these are more massive and carry more momentum, their de Broglie wavelength is even smaller, which gives improved resolution.

AS1 Particles, Radiation and Quantum Phenomena Sample module test

1 (a) Describe the principal features of the nuclear model of the atom suggested by Rutherford.

....
....
....
....
.... *(5)*

(b) When gold foil is bombarded by α particles it is found that most of the particles pass through the foil without significant change of direction or loss of energy. A few particles are deviated from their original direction by more than 90°. Explain, in terms of the nuclear model of the atom and by considering the nature of the forces acting,

  (i) why some α particles are deflected through large angles,

....
....
....
....

  (ii) why most of the particles pass through the foil without significant change in direction.

....
....
....
....
.... *(5)*
                                                                                                                                          (10)

2 A particle accelerator is designed to accelerate antiprotons through a potential difference of 2.0 GV and make them collide with protons of equal energy moving in the opposite direction. In such a collision, a proton–antiproton pair is created as represented by the equation

$$p + \bar{p} \rightarrow p + \bar{p} + p + \bar{p}$$

You may assume that the rest energy of the proton is 940 MeV.

(a) State how an antiproton differs from a proton.

.... *(1)*

(b) Give the total kinetic energy of the particles, in GeV, before collision.

.... *(1)*

(c) State the rest energy of the antiproton.

.... *(1)*

(d) Calculate the total kinetic energy of the particles, in GeV, after the collision.

....

....

....

.... *(3)*

(6)

3 (a) State which interaction, strong or weak, is experienced by each of the following particles.

hadrons .... *(1)*

leptons .... *(1)*

(b) Give one example of a hadron and one example of a lepton.

hadron .... *(1)*

lepton .... *(1)*

(c) Hadrons are classified as either baryons or mesons. How many quarks are there in a baryon and in a meson?

baryon .... *(1)*

meson .... *(1)*

(d) (i) State the quark composition of a neutron. ....

(ii) Describe, in terms of quarks, the process of β⁻ decay when a neutron changes into a proton.

....

....

....

(iii) Sketch a Feynman diagram to represent β⁻ decay. ....

(4)

(10)

4 (a) State Snell's law of refraction of light and explain the conditions under which the law applies.

....

....

.... *(3)*

(b) The diagram shows a glass pentaprism as used in the viewfinder of some cameras. Light enters face AB and leaves face BC. The faces AE, ED and DC are silvered and the refractive index of the glass is 1.52.

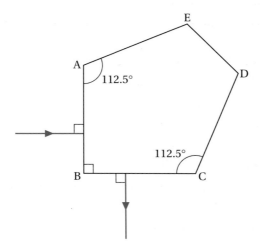

(i) On the diagram, draw the path of the incident ray from face AB to CD.

(ii) State why you have drawn the ray in this direction.

....

.... (2)

(c) Explain, with the aid of a calculation, why the face CD needs to be silvered if the ray shown is not to be refracted at face CD.

....

....

.... (3)

(d) On the diagram, continue the ray until it leaves the prism. *(1)*

(9)

5 (a) Calculate the wavelength of a γ-ray photon which has an energy of 1.6×10^{-15} J.

....

.... (2)

(b) An X-ray photon is generated which has the same energy as the γ-ray photon described in part (a).

(i) How do the speeds in a vacuum of these two photons compare?

.... (1)

(ii) How does their ability to penetrate a given material compare?

.... (1)

(4)

6 (a) In the photoelectric effect equation $hf = \phi + E_k$ state what is meant by

hf ....

ϕ ....

E_k .... (3)

(b) Monochromatic light of wavelength 3.80×10^{-7} m falls with an intensity of 6.0 µW m^{-2} on to a metallic surface whose work function is 3.2×10^{-19} J. The Planck constant is 6.6×10^{-34} J s and the speed of light is 3.0×10^{8} m s^{-1}. Calculate

  (i) the energy of a single photon of light of this wavelength,

  ....
  ....
  ....

  (ii) the number of electrons emitted per second from 1.0×10^{-6} m^2 of the surface if the photon has a 1 in 1000 chance of ejecting an electron,

  ....
  ....
  ....
  ....

  (iii) the maximum kinetic energy which one of these photoelectrons could possess.

  ....
  .... (5)
  .... (8)

7 (a) (i) Explain what is meant by the *duality of electrons*.

  ....
  ....

  (ii) State the relation between the electron mass, the electron velocity and the wavelength for a monoenergetic beam of electrons.

  ....
  .... (3)

(b) The spacing of atoms in a crystal is 1.0×10^{-10} m. The mass of the electron is 9.1×10^{-31} kg and the Planck constant is 6.6×10^{-34} J s. Estimate the speed of electrons which would give detectable diffraction effects with such crystals.

  ....
  ....
  .... (4)

(c) Give one piece of evidence to demonstrate that electrons have particle properties.

  ....
  .... (1)
  .... (8)

Module test answers

The sign / indicates alternative answers, either of which will gain you the mark available.

1 (a) nucleus is small   1
    nucleus is massive   1
    nucleus is positive   1
    electrons surround the nucleus   1
    electrons are negative / electrons have small mass   1

(b) (i) some α particles approach (gold) nucleus closely 1
    (gold) nucleus and α particles are *both* positively charged   1
    repulsive force between (gold) nucleus and α particle   1

(ii) space between (gold) nuclei is large / atom is mostly empty space   1
    most α particles do not approach nuclei close enough to be (significantly) deflected   1

2 (a) proton is positively charged, antiproton is negatively charged   1

(b) 4.0 (GeV)   1

(c) rest energy of antiproton is also 940 MeV   1

(d) energy needed to create proton–antiproton pair
    $= 2 \times 940$ MeV   1
    $= 1.9$ GeV   1
    kinetic energy remaining $= 4.0 - 1.9 = 2.1$ (GeV)   1

3 (a) hadrons interact through the strong interaction   1
    leptons interact through the weak interaction   1

(b) hadron example, e.g. proton, neutron   1
    lepton example, e.g. electron, neutrino   1

(c) baryon = 3 quarks / antiquarks   1
    meson = quark + antiquark   1

(d) (i) udd   1
    (ii) d changes to u   1
    (iii) W^- emitted
    W^- decays to electron and antineutrino
    both shown correctly as on diagram below   2

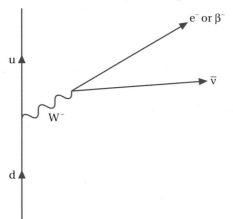

4 (a) for monochromatic light   1
    travelling from one medium to another   1
    $\dfrac{\sin(\text{angle of incidence})}{\sin(\text{angle of refraction})} = $ constant   1

(b)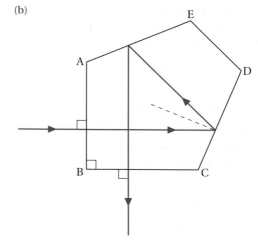

(i) correct ray as in diagram above   1
(ii) no deviation because $i = 0$ / because ray is normal to AB   1

(c) $\sin \theta_c = \dfrac{1}{1.52}$ $i_c = 41°$ 1
    $\theta = 22.5°$ 1
    $\theta < i_c$, refracted angle $= 35.6°$, so no total internal reflection / rays would emerge   1

(d) correct ray as on diagram above   1

5 (a) $\lambda = \left(\dfrac{hc}{E}\right) = \dfrac{6.63 \times 10^{-34} \times 3 \times 10^8}{1.6 \times 10^{-15}}$ 1
    $= 1.2(4) \times 10^{-10}$ m   1

(b) (i) same   1
    (ii) same   1

6 (a) $hf =$ photon energy   1
    $\phi =$ work function   1
    $E_k =$ maximum kinetic energy of photoelectrons   1

(b) (i) $E = \left(\dfrac{hc}{\lambda}\right) = \dfrac{6.6 \times 10^{-34} \times 3 \times 10^8}{3.8 \times 10^{-7}} = 5.23 \times 10^{-19}$ J   1

(ii) energy on surface $= 6.0 \times 10^{-12}$ J s^{-1} 1
    $N_p = \dfrac{6.0 \times 10^{-12}}{5.23 \times 10^{-19}} = 1.1(5) \times 10^7$ s^{-1} 1
    but only 1 in 1000 will release an electron, so $N_e = 1.1 \times 10^4$ s^{-1}

(iii) $E_k = \dfrac{hc}{\lambda} - \phi = (5.2(3) - 3.2) \times 10^{-19} = 2.0 \times 10^{-19}$ J   2

7 (a) (i) electrons behave sometimes as particles   1
        and sometimes as waves   1
   (ii) $mv \propto 1/\lambda$ or $mv = h/\lambda$ 1

(b) for diffraction, electron wavelength must be of order of atom spacing   1
    hence $\lambda = 10^{-10}$ m   1
    $v = \dfrac{h}{m\lambda}$ 1

    $= \dfrac{6.6 \times 10^{-34}}{9.1 \times 10^{-31} \times 10^{-10}} = 7.2(5) \times 10^{6}$ m s^{-1} 1

(c) deflection in electric field (or deflection in magnetic field, or any other correct evidence)   1